# REMEMBER TO LOOK UP AT THE STARS AND NOT DOWN AT YOUR FEET

T0187313

# How Did It All Begin?

**STEPHEN HAWKING** was a brilliant theoretical physicist and is generally considered to have been one of the world's greatest thinkers. He held the position of Lucasian Professor of Mathematics at the University of Cambridge for thirty years and is the author of *A Brief History of Time*, which was an international bestseller. His other books for the general reader include *A Briefer History of Time*, the essay collection *Black Holes and Baby Universes*, *The Universe in a Nutshell*, *The Grand Design* and *Black Holes: The BBC Reith Lectures*. He died on 14 March 2018.

'How did it all begin?' and 'Is there a God' are essays taken from Stephen Hawking's final book, *Brief Answers to the Big Questions* (John Murray, 2018).

# STEPHEN HAWKING

How Did It All Begin?

JOHN MURRAY

First published in Great Britain in 2022 by John Murray (Publishers)
An Hachette UK company

1

'How did it all begin?' and 'Is there a God?' are essays taken from *Brief
Answers to the Big Questions*, published by John Murray (2018)

A CIP catalogue record for this title
is available from the British Library

Paperback ISBN 978-1-529-39242-5
eBook ISBN 978-1-529-39243-2

Text design by Craig Burgess

Typeset in Sabon MT by
Palimpsest Book Production Ltd, Falkirk, Stirlingshire

Printed and bound in Great Britain by Clays Ltd, Elcograf S.p.A.

John Murray policy is to use papers that are natural, renewable and
recyclable products and made from wood grown in sustainable forests.
The logging and manufacturing processes are expected to conform
to the environmental regulations of the country of origin.

John Murray (Publishers)
Carmelite House
50 Victoria Embankment
London EC4Y 0DZ

www.johnmurraypress.co.uk

# Contents

# HOW DID IT ALL BEGIN?

HAMLET said, 'I could be bounded in a nutshell, and count myself a king of infinite space.' I think what he meant was that although we humans are very limited physically, particularly in my own case, our minds are free to explore the whole universe, and to boldly go where even *Star Trek* fears to tread. Is the universe actually infinite, or just very large? Did it have a beginning? Will it last for ever or just a long time? How can our finite minds comprehend an infinite universe? Isn't it pretentious of us even to make the attempt?

At the risk of incurring the fate of Prometheus, who stole fire from the ancient gods for human use, I believe we can, and should, try to understand the universe. Prometheus' punishment was being chained to a rock for eternity, although happily he was eventually liberated by Hercules. We have already made remarkable progress in understanding the cosmos. We don't yet have a complete picture. I like to think we may not be far off.

According to the Boshongo people of central Africa, in the beginning there was only darkness, water and the great god Bumba. One day Bumba, in pain from stomach ache, vomited up the Sun. The Sun dried up some of the water, leaving land. Still in pain, Bumba vomited up the Moon, the stars and then some animals – the leopard, the crocodile, the turtle and, finally man.

These creation myths, like many others,

try to answer the questions we all ask. Why are we here? Where did we come from? The answer generally given was that humans were of comparatively recent origin because it must have been obvious that the human race was improving its knowledge and technology. So it can't have been around that long or it would have progressed even more. For example, according to Bishop Ussher, the Book of Genesis placed the beginning of time on 22 October 4004 BCE at 6 p.m. On the other hand, the physical surroundings, like mountains and rivers, change very little in a human lifetime. They were therefore thought to be a constant background, and either to have existed for ever as an empty landscape, or to have been created at the same time as the humans.

Not everyone however was happy with

the idea that the universe had a beginning. For example, Aristotle, the most famous of the Greek philosophers, believed that the universe had existed for ever. Something eternal is more perfect than something created. He suggested the reason we see progress was that floods, or other natural disasters, had repeatedly set civilisation back to the beginning. The motivation for believing in an eternal universe was the desire to avoid invoking divine intervention to create the universe and set it going. Conversely, those who believed that the universe had a beginning used it as an argument for the existence of God as the first cause, or prime mover, of the universe.

If one believed that the universe had a beginning, the obvious questions were, 'What happened before the beginning?

What was God doing before he made the world? Was he preparing Hell for people who asked such questions?' The problem of whether or not the universe had a beginning was a great concern to the German philosopher Immanuel Kant. He felt there were logical contradictions, or antimonies, either way. If the universe had a beginning, why did it wait an infinite time before it began? He called that the thesis. On the other hand, if the universe had existed for ever, why did it take an infinite time to reach the present stage? He called that the antithesis. Both the thesis and the antithesis depended on Kant's assumption, along with almost everyone else, that time was absolute. That is to say, it went from the infinite past to the infinite future independently of any universe that might or might not exist.

This is still the picture in the mind of many scientists today. However, in 1915 Einstein introduced his revolutionary general theory of relativity. In this, space and time were no longer absolute, no longer a fixed background to events. Instead, they were dynamical quantities that were shaped by the matter and energy in the universe. They were defined only within the universe, so it made no sense to talk of a time before the universe began. It would be like asking for a point south of the South Pole. It is not defined.

Although Einstein's theory unified time and space, it didn't tell us much about space itself. Something that seems obvious about space is that it goes on and on and on. We don't expect the universe to end in a brick wall, although there's no logical reason why it couldn't. But modern instruments like the Hubble space telescope allow us to probe deep into space. What we see is billions and billions of galaxies, of various shapes and sizes. There are giant elliptical galaxies, and spiral galaxies like our own. Each galaxy contains billions and billions of stars, many of which will have planets round them. Our own galaxy blocks our view in certain directions, but apart from that the galaxies are distributed roughly uniformly throughout space, with some local concentrations and voids. The density of galaxies appears to drop off at very large distances, but that seems to be because they are so far away and faint that we can't make them out. As far as we can tell, the universe goes on in space for ever and is much the same no matter how far it goes on.

Although the universe seems to be

much the same at each position in space, it is definitely changing in time. This was not realised until the early years of the last century. Up to then, it was thought the universe was essentially constant in time. It might have existed for an infinite time, but that seemed to lead to absurd conclusions. If stars had been radiating for an infinite time, they would have heated up the universe until it reached their own temperature. Even at night, the whole sky would be as bright as the Sun, because every line of sight would have ended either on a star or on a cloud of dust that had been heated up until it was as hot as the stars. So the observation that we have all made, that the sky at night is dark, is very important. It implies that the universe cannot have existed for ever, in the state we see today. Something must have happened in the past to make the stars turn on a finite time ago. Then the light from very distant stars wouldn't have had time to reach us yet. This would explain why the sky at night isn't glowing in every direction.

If the stars had just been sitting there for ever, why did they suddenly light up a few billion years ago? What was the clock that told them it was time to shine? This puzzled those philosophers, like Immanuel Kant, who believed that the universe had existed for ever. But for most people it was consistent with the idea that the universe had been created, much as it is now, only a few thousand years ago, just as Bishop Ussher had concluded. However, discrepancies in this idea began to appear, with observations by the hundred-inch telescope on Mount Wilson in the 1920s. First of all, Edwin Hubble discovered that many faint patches of light, called nebulae, were in fact other galaxies, vast collections of stars like our Sun, but at a great distance. In order for them to appear so small and faint, the distances had to be so great that light from them would have taken millions or even billions of years, to reach us. This indicated that the beginning of the universe couldn't have been just a few thousand years ago.

But the second thing Hubble discovered was even more remarkable. By an analysis of the light from other galaxies, Hubble was able to measure whether they were moving towards us or away. To his great surprise, he found they were nearly all moving away. Moreover, the further they were from us, the faster they were moving away. In other words, the universe is expanding. Galaxies are moving away from each other.

The discovery of the expansion of the universe was one of the great intellectual revolutions of the twentieth century. It came as a total surprise, and it completely changed the discussion of the origin of the universe. If the galaxies are moving apart, they must have been closer together in the past. From the present rate of expansion, we can estimate that they must have been very close together indeed, about 10 to 15 billion years ago. So it looks as though the universe might have started then, with everything being at the same point in space.

But many scientists were unhappy with the universe having a beginning, because it seemed to imply that physics broke down.

One would have to invoke an outside agency, which for convenience one can call God, to determine how the universe began. They therefore advanced theories in which the universe was expanding at the present time, but didn't have a beginning. One of these was the steady-state theory, proposed by Hermann Bondi, Thomas Gold and Fred Hoyle in 1948.

In the steady-state theory, as galaxies moved apart, the idea was that new galaxies would form from matter that was supposed to be continually being created throughout space. The universe would have existed for ever, and would have looked the same at all times. This last property had the great virtue of being a definite prediction that could be tested by observation. The Cambridge radio astronomy group, under Martin Ryle, did a survey of weak sources of radio waves in the early 1960s. These were distributed fairly uniformly across the sky, indicating that most of the sources lay outside our galaxy. The weaker sources would be further away, on average.

The steady-state theory predicted a relationship between the number of sources and their strength. But the observations showed more faint sources than predicted, indicating that the density of the sources was higher in the past. This was contrary to the basic assumption of the steady-state theory, that everything was constant in time. For this and other reasons, the steady-state theory was abandoned.

Another attempt to avoid the universe having a beginning was the suggestion that there was a previous contracting phase, but because of rotation and local irregularities the matter would not all fall to the same point. Instead, different parts of the matter would miss each other, and the universe would expand again with the density always remaining finite. Two Russians, Lifshitz and Khalatnikov, actually claimed to have proved that a general contraction without exact symmetry would always lead to a bounce, with the density remaining finite. This result was very convenient for Marxist-Leninist dialectical materialism, because it avoided awkward questions about the creation of the universe. It therefore became an article of faith for Soviet scientists.

I began my research in cosmology just about the time that Lifshitz and Khalatnikov published their conclusion that the universe didn't have a beginning. I realised that this was a very important question, but I wasn't convinced by the arguments that Lifshitz and Khalatnikov had used.

We are used to the idea that events are caused by earlier events, which in turn are caused by still earlier events. There is a chain of causality, stretching back into the past. But suppose this chain has a beginning, suppose there was a first event. What caused it? This was not a question that many scientists wanted to address. They tried to avoid it, either by claiming like the Russians and the steady-state theorists that the universe didn't have a beginning or by maintaining that the origin of the universe did not lie within the realm of science but belonged to metaphysics or religion. In my opinion, this is not a position any true scientist should take. If the laws of science are suspended at the beginning of the universe, might not they also fail at other times? A law is not a law if it only holds sometimes. I believe that we should try to understand the beginning of the universe on the basis of science. It may be a task beyond our powers, but at least we should make the attempt.

Roger Penrose and I managed to prove geometrical theorems to show that the universe must have had a beginning if Einstein's general theory of relativity was correct, and certain reasonable conditions were satisfied. It is difficult to argue with a mathematical theorem, so in the end Lifshitz and Khalatnikov conceded that the universe should have a beginning. Although the idea of a beginning to the universe might not be very welcome to communist ideas, ideology was never allowed to stand in the way of science in physics. Physics was needed for the bomb, and it was important that it worked. However, Soviet ideology did prevent progress in biology by denying the truth of genetics.

Although the theorems Roger Penrose and I proved showed that the universe must have had a beginning, they did not give much information about the nature of that beginning. They indicated that the universe began in a Big Bang, a point where the whole universe and everything in it were scrunched up into a single point of infinite density, a space-time singularity. At this point Einstein's general theory of relativity would have broken down. Thus one cannot use it to predict in what manner the universe began. One is left with the origin of the universe apparently being beyond the scope of science.

Observational evidence to confirm the idea that the universe had a very dense beginning came in October 1965, a few months after my first singularity result, with the discovery of a faint background of microwaves throughout space. These microwaves are the same as those in your microwave oven, but very much less powerful. They would heat your pizza only to minus 270.4 degrees centigrade, not much good for defrosting the pizza, let alone cooking it. You can actually observe these microwaves yourself. Those of you who remember analogue televisions have almost certainly observed these microwaves. If you ever set your television to an empty channel, a few per cent of the snow you saw on the screen was caused by this background of microwaves. The only reasonable interpretation of the background is that it is radiation left over from an early very hot and dense state. As the universe expanded, the radiation would have cooled until it is just the faint remnant we observe today.

That the universe began with a singularity was not an idea that I or a number of other people were happy with. The reason Einstein's general relativity breaks down near the Big Bang is that it is what is called a classical theory. That is, it implicitly assumed what seems obvious from common sense, that each particle had a well-defined position and a well-defined speed. In such a so-called classical theory, if one knows the positions and speeds of all the particles in the universe at one time, one can calculate what they would be at any other time, in the past or future. However, in the early twentieth century scientists discovered that they couldn't calculate exactly what would happen over very short distances. It wasn't just that they needed better theories. There seems to be a certain level of randomness or uncertainty in nature that cannot be removed however good our theories. It can be summed up in the Uncertainty Principle that was formulated in 1925 by the German scientist Werner Heisenberg. One cannot accurately predict both the position and the speed of a particle. The more accurately the position is predicted, the less accurately you will be able to predict the speed, and vice versa.

Einstein objected strongly to the idea that the universe is governed by chance. His feelings were summed up in his dictum 'God does not play dice.' But all the evidence is that God is quite a gambler. The universe is like a giant casino with dice being rolled, or wheels being spun, on every occasion. A casino owner risks losing money each time dice are thrown or the roulette wheel is spun. But over a large number of bets the odds average out, and the casino owner makes sure they average out in his or her favour. That's why casino owners are so rich. The only chance you have of winning against them is to stake all your money on a few rolls of the dice or spins of the wheel.

It is the same with the universe. When the universe is big, there are a very large number of rolls of the dice, and the results average out to something one can predict. But when the universe is very small, near the Big Bang, there are only a small number of rolls of the dice, and the Uncertainty

Principle is very important. In order to understand the origin of the universe, one therefore has to incorporate the Uncertainty Principle into Einstein's general theory of relativity. This has been the great challenge in theoretical physics for at least the last thirty years. We haven't solved it yet, but we have made a lot of progress.

Now suppose we try to predict the future. Because we only know some combination of position and speed of a particle, we cannot make precise predictions about the future positions and speeds of particles. We can only assign a probability to particular combinations of positions and speeds. Thus there is a certain probability to a particular future of the universe. But now suppose we try to understand the past in the same way.

Given the nature of the observations we can make now, all we can do is assign a probability to a particular history of the universe. Thus the universe must have many possible histories, each with its own probability. There is a history of the universe in which England win the World Cup again, though maybe the probability is low. This idea that the universe has multiple histories may sound like science fiction, but it is now accepted as science fact. It is due to Richard Feynman, who worked at the eminently respectable California Institute of Technology and played the bongo drums in a strip joint up the road. The way Feynman's approach to understanding how things works is to assign to each possible history a particular probability, and then use this idea to make predictions. It works spectacularly well to predict the future. So we presume it works to retrodict the past too.

Scientists are now working to combine Einstein's general theory of relativity and Feynman's idea of multiple histories into a complete unified theory that will describe everything that happens in the universe. This unified theory will enable us to calculate how the universe will evolve, if we know its state at one time. But the unified theory will not in itself tell us how the universe began, or what its initial state was. For that, we need something extra. We require what are known as boundary conditions, things that tell us what happens at the frontiers of the universe, the edges of space and time. But if the frontier of the universe was just at a normal point of space and time we could go past it and claim the territory beyond as part of the universe. On the other hand, if the boundary of the universe was at a jagged edge where space or time were scrunched up, and the density was infinite, it would be very difficult to define meaningful boundary conditions. So it is not clear what boundary conditions are needed. It seems there is no logical basis for picking one set of boundary conditions over another.

However, Jim Hartle of the University of California, Santa Barbara, and I realised there was a third possibility. Maybe the universe has no boundary in space and time. At first sight, this seems to be in direct contradiction to the geometrical theorems that I mentioned earlier. These showed that the universe must have had a beginning, a boundary in time. However, in order to make Feynman's techniques mathematically well defined, the mathematicians developed a concept called imaginary time. It isn't anything to do with the real time that we experience. It is a mathematical trick to make the calculations work and it replaces the real time we experience. Our idea was to say that there was no boundary in imaginary time. That did away with trying to invent boundary conditions. We called this the no-boundary proposal.

If the boundary condition of the universe is that it has no boundary in imaginary time, it won't have just a single history. There are many histories in imaginary time and each of them will determine a history in real time. Thus we have a superabundance of histories for the universe. What picks out the particular history, or set of histories that we live in, from the set of all possible histories of the universe?

One point that we can quickly notice is that many of these possible histories of the universe won't go through the sequence of forming galaxies and stars, something that was essential to our own development. It may be that intelligent beings can evolve without galaxies and stars, but it seems unlikely. Thus the very fact that we exist as beings that can ask the question 'Why is the universe the way it is?' is a restriction on the history we live in. It implies it is one of the minority of histories that have galaxies and stars. This is an example of what is called the Anthropic Principle. The Anthropic Principle says that the universe has to be more or less as we see it, because if it were different there wouldn't be anyone here to observe it.

Many scientists dislike the Anthropic Principle, because it seems little more than hand waving, and not to have much predictive power. But the Anthropic Principle can be given a precise formulation, and it seems to be essential when dealing with the origin of the universe. M-theory, which is our best candidate for a complete unified theory, allows a very large number of possible histories for the universe. Most of these histories are quite unsuitable for the development of intelligent life. Either they are empty, or too short lasting, or too highly curved, or wrong in some other way. Yet, according to Richard Feynman's multiple-histories idea, these uninhabited histories might have quite a high probability.

We really don't care how many histories there may be that don't contain intelligent beings. We are interested only in the subset of histories in which intelligent life develops. This intelligent life need not be anything like humans. Little green men would do as well. In fact, they might do rather better. The human race does not have a very good record of intelligent behaviour.

As an example of the power of the Anthropic Principle, consider the number of directions in space. It is a matter of common experience that we live in three-dimensional space. That is to say, we can represent the position of a point in space by three numbers. For example, latitude, longitude and height above sea level. But why is space three-dimensional? Why isn't it two, or four, or some other number of dimensions, like in science fiction? In fact in M-theory space has ten dimensions, but it is thought that seven of the directions are curled up very small, leaving three directions that are large and nearly flat. It is like a drinking straw. The surface of a straw is two-dimensional. However, one direction is curled up into a small circle, so that from a distance the straw looks like a one-dimensional line.

Why don't we live in a history in which eight of the dimensions are curled up small, leaving only two dimensions that we notice? A two-dimensional animal would have a hard job digesting food. If it had a gut that went right through, like we have, it would divide the animal in two, and the poor creature would fall apart. So two flat directions are not enough for anything as complicated as intelligent life. There is something special about three space dimensions. In three dimensions, planets can have stable orbits around stars. This is a consequence of gravitation obeying the inverse square law, as discovered by Robert Hooke in 1665 and elaborated on by Isaac Newton. Think about the gravitational attraction of two bodies at a particular distance. If that distance is doubled, then the force between them is halved. If the distance is tripled then the force is divided by nine, if quadrupled, then the force is divided by sixteen and so on. This leads to stable planetary orbits. Now let's think about four space dimensions. There gravitation would obey an inverse cube law. If the distance between two bodies is doubled, then the gravitational force would be divided by eight, tripled by twenty-seven and if quadrupled, by sixty-four. This change to an inverse cube law prevents planets from having stable orbits around their suns. They would either fall into their sun or escape to the outer darkness and cold. Similarly, the orbits of electrons in atoms would not be stable, so matter as we know it would not exist. Thus although the multiple-histories idea would allow any number of nearly flat directions, only histories with three flat directions will contain intelligent beings. Only in such histories will the question be asked, 'Why does space have three dimensions?'

One remarkable feature of the universe we observe concerns the microwave background discovered by Arno Penzias and Robert Wilson. It is essentially a fossil record of how the universe was when very young. This background is almost the same independently of which direction one looks in. The differences between different directions is about one part in 100,000. These differences are incredibly tiny and need an explanation. The generally accepted explanation for this smoothness is that very early in the history of the universe it underwent a period of very rapid expansion, by a factor of at least a billion billion billion. This process is known as inflation, something that was good for the universe in contrast to inflation of prices that too often plagues us. If that was all there was to it, the microwave radiation would be totally the same in all directions. So where did the small discrepancies come from?

In early 1982, I wrote a paper proposing that these differences arose from the quantum fluctuations during the inflationary period. Quantum fluctuations occur as a consequence of the Uncertainty

Principle. Furthermore, these fluctuations were the seeds for structures in our universe: galaxies, stars and us. This idea is basically the same mechanism as so-called Hawking radiation from a black hole horizon, which I had predicted a decade earlier, except that now it comes from a cosmological horizon, the surface that divided the universe between the parts that we can see and the parts that we cannot observe. We held a workshop in Cambridge that summer, attended by all the major players in the field. At this meeting, we established most of our present picture of inflation, including the all-important density fluctuations, which give rise to galaxy formation and so to our existence. Several people contributed to the final answer. This was ten years before fluctuations in the microwave sky were discovered by the COBE satellite in 1993, so theory was way ahead of experiment.

Cosmology became a precision science another ten years later, in 2003, with the first results from the WMAP satellite. WMAP produced a wonderful map of the temperature of the cosmic microwave sky, a snapshot of the universe at about one-hundredth of its present age. The irregularities you see are predicted by inflation, and they mean that some regions of the universe had a slightly higher density than others. The gravitational attraction of the extra density slows the expansion of that region, and can eventually cause it to collapse to form galaxies and stars. So look carefully at the map of the microwave sky. It is the blueprint for all the structure in the universe. We are the product of quantum fluctuations in the very early universe. God really does play dice.

Superseding WMAP, today there is the Planck satellite, with a much higher-resolution map of the universe. Planck is testing our theories in earnest, and may even detect the imprint of gravitational waves predicted by inflation. This would be quantum gravity written across the sky.

There may be other universes. M-theory predicts that a great many universes were created out of nothing, corresponding to the many different possible histories. Each universe has many possible histories and many possible states as they age to the present and beyond into the future. Most of these states will be quite unlike the universe we observe.

There is still hope that we see the first evidence for M-theory at the LHC particle accelerator, the Large Hadron Collider, at CERN in Geneva. From an M-theory perspective, it only probes low energies, but we might be lucky and see a weaker signal of fundamental theory, such as supersymmetry. I think the discovery of supersymmetric partners for the known particles would revolutionise our understanding of the universe.

In 2012, the discovery of the Higgs particle by the LHC at CERN in Geneva was announced. This was the first discovery of a new elementary particle in the twenty-first century. There is still some hope that the LHC will discover supersymmetry. But even if the LHC does not discover any new elementary particles, supersymmetry might still be found in the next generation of accelerators that are presently being planned.

The beginning of the universe itself in the Hot Big Bang is the ultimate high-energy laboratory for testing M-theory, and our ideas about the building blocks of space-time and matter. Different theories leave behind different fingerprints in the current structure of the universe, so astrophysical data can give us clues about the unification of all the forces of nature. So there may well be other universes, but unfortunately we will never be able to explore them.

We have seen something about the origin of the universe. But that leaves two big questions. Will the universe end? Is the universe unique?

What then will be the future behaviour of the most probable histories of the universe? There seem to be various possibilities, which are compatible with the appearance of intelligent beings. They depend on the amount of matter in the universe. If there is more than a certain critical amount, the gravitational attraction between the galaxies will slow down the expansion.

Eventually they will then start falling towards each other and will all come together in a Big Crunch. That will be the end of the history of the universe, in real time. When I was in the Far East, I was asked not to mention the Big Crunch, because of the effect it might have on the market. But the markets crashed, so maybe the story got out somehow. In Britain, people don't seem too worried about a possible end twenty billion years in the future. You can do quite a lot of eating, drinking and being merry before that.

If the density of the universe is below the critical value, gravity is too weak to stop the galaxies flying apart for ever. All the stars will burn out, and the universe will get emptier and emptier, and colder and colder. So, again, things will come to an end, but in a less dramatic way. Still, we have a few billion years in hand.

In this answer, I have tried to explain something of the origins, future and nature of our universe. The universe in the past was small and dense and so it is quite like the nutshell with which I began. Yet this nut encodes everything that happens in real time. So Hamlet was quite right. We could be bounded in a nutshell and count ourselves kings of infinite space.

I REGARD IT A TRIUMPH THAT WE, WHO ARE OURSELVES MERE STARDUST, HAVE COME TO SUCH A DETAILED UNDERSTANDING OF THE UNIVERSE IN WHICH WE LIVE

# IS THERE A GOD?

SCIENCE is increasingly answering questions that used to be the province of religion. Religion was an early attempt to answer the questions we all ask: why are we here, where did we come from? Long ago, the answer was almost always the same: gods made everything. The world was a scary place, so even people as tough as the Vikings believed in supernatural beings to make sense of natural phenomena like lightning, storms or eclipses. Nowadays, science provides better and more consistent answers, but people will always cling to religion, because it gives comfort, and they do not trust or understand science.

A few years ago, *The Times* newspaper ran a headline on the front page which said '"There is no God," says Hawking'. The article was illustrated. God was shown in a drawing by Leonardo da Vinci, looking thunderous. They printed a photo of me, looking smug. They made it look like a duel between us. But I don't have a grudge against God. I do not want to give the impression that my work is about proving or disproving the existence of God. My work is about finding a rational framework to understand the universe around us.

For centuries, it was believed that disabled people like me were living under a curse that was inflicted by God. Well, I suppose it's possible that I've upset someone up there, but I prefer to think that everything can be explained another way, by the laws of nature. If you believe in science, like I do, you believe that there are certain laws that are always obeyed. If you like, you can say the laws are the work of God, but that is more a definition of God than a proof of his existence. In about 300 BCE, a philosopher called Aristarchus was fascinated by eclipses, especially eclipses of the Moon. He was brave enough to question whether they really were caused by gods. Aristarchus was a true scientific pioneer. He studied the heavens carefully and reached a bold conclusion: he realised the eclipse was really the shadow of the Earth passing over the Moon, and not a divine event. Liberated by this discovery, he was able to work out what was really going on above his head, and draw diagrams that showed the true relationship of the Sun, the Earth and the Moon. From there he reached even more remarkable conclusions. He deduced that the Earth was not the centre of the universe, as everyone had thought, but that it instead orbits the Sun. In fact, understanding this arrangement explains all eclipses. When the Moon casts its shadow on the Earth, that's a solar eclipse. And when the Earth shades the Moon, that's a lunar eclipse. But Aristarchus took it even further. He suggested that stars were not chinks in the floor of heaven, as his contemporaries believed, but that stars were other suns, like ours, only a very long way away. What a stunning realisation it must have been. The universe is a machine governed by principles or laws – laws that can be understood by the human mind.

I believe that the discovery of these laws has been humankind's greatest achievement, for it's these laws of nature – as we now call them – that will tell us whether we need a god to explain the universe at all. The laws of nature are a description of how things actually work in the past, present and future. In tennis, the ball always goes exactly where they say it will. And there are many other laws at work here too. They govern everything that is going on, from how the energy of the shot is produced in the players' muscles to the speed at which the grass grows beneath their feet. But what's really important is that these physical laws, as well as being unchangeable, are universal. They apply not just to the flight of a ball, but to the motion of a planet, and everything else in the universe. Unlike laws made by humans, the laws of nature cannot be broken – that's why they are so powerful and, when seen from a religious standpoint, controversial too.

If you accept, as I do, that the laws of nature are fixed, then it doesn't take long to ask: what role is there for God? This is a big part of the contradiction between science and religion, and although my views have made headlines, it is actually an ancient conflict. One could define God as the embodiment of the laws of nature. However, this is not what most people would think of as God. They mean a human-like being, with whom one can have a personal relationship. When you look at the vast size of the universe, and how insignificant and accidental human life is in it, that seems most implausible.

I use the word 'God' in an impersonal sense, like Einstein did, for the laws of nature, so knowing the mind of God is knowing the laws of nature. My prediction is we will know the mind of God by the end of this century.

The one remaining area that religion can now lay claim to is the origin of the universe, but even here science is making progress and should soon provide a definitive answer to how the universe began. I published a book that asked if God created the universe, and that caused something of a stir. People got upset that a scientist should have anything to say on the matter of religion. I have no desire to tell anyone what to believe, but for me asking if God exists is a valid question for science. After all, it is hard to think of a more important, or fundamental, mystery than what, or who, created and controls the universe.

I think the universe was spontaneously created out of nothing, according to the laws of science. It has no beginning, and no end. The basic assumption of science is scientific determinism. The laws of science determine the evolution of the universe, given its state at one time. These laws may, or may not, have been decreed by God, but he cannot intervene to break the laws, or they would not be laws. That leaves God with the freedom to choose the initial state of the universe, but even here it seems there may be laws. So God would have no freedom at all.

Despite the complexity and variety of the universe, it turns out that to make one you need just three ingredients. Let's imagine that we could list them in some kind of cosmic cookbook. So what are the three ingredients we need to cook up a universe? The first is matter – stuff that has mass. Matter is all around us, in the ground beneath our feet and out in space.

Dust, rock, ice, liquids. Vast clouds of gas, massive spirals of stars, each containing billions of suns, stretching away for incredible distances.

The second thing you need is energy. Even if you've never thought about it, we all know what energy is. Something we encounter every day. Look up at the Sun and you can feel it on your face: energy produced by a star ninety-three million miles away. Energy permeates the universe, driving the processes that keep it a dynamic, endlessly changing place.

So we have matter and we have energy. The third thing we need to build a universe is space. Lots of space. You can call the universe many things – awesome, beautiful, violent – but one thing you can't call it is cramped. Wherever we look we see space, more space and even more space.

Stretching in all directions. It's enough to make your head spin. So where could all this matter, energy and space come from? We had no idea until the twentieth century.

The answer came from the insights of one man, probably the most remarkable scientist who has ever lived. His name was Albert Einstein. Sadly I never got to meet him, since I was only thirteen when he died. Einstein realised something quite remarkable: that two of the main ingredients needed to make a universe – mass and energy – are basically the same thing, two sides of the same coin if you like. His famous equation $E = mc^2$ simply means that mass can be thought of as a kind of energy, and vice versa. So instead of three ingredients, we can now say that the universe has just two: energy and space. So where did all this energy and space come from? The answer was found after decades of work by scientists: space and energy were spontaneously invented in an event we now call the Big Bang.

At the moment of the Big Bang, an entire universe came into existence, and with it space. It all inflated, just like a balloon being blown up. So where did all this energy and space come from? How does an entire universe full of energy, the awesome vastness of space and everything in it, simply appear out of nothing?

For some, this is where God comes back into the picture. It was God who created the energy and space. The Big Bang was the moment of creation. But science tells a different story. At the risk of getting myself into trouble, I think we can understand much more the natural phenomena that terrified the Vikings. We can even go beyond the beautiful symmetry of energy and matter discovered by Einstein. We can use the laws of nature to address the very origins of the universe, and discover if the existence of God is the only way to explain it.

As I was growing up in England after the Second World War, it was a time of austerity. We were told that you never get something for nothing. But now, after a lifetime of work, I think that actually you can get a whole universe for free.

The great mystery at the heart of the Big Bang is to explain how an entire, fantastically enormous universe of space and energy can materialise out of nothing. The secret lies in one of the strangest facts about our cosmos. The laws of physics demand the existence of something called 'negative energy'.

To help you get your head around this weird but crucial concept, let me draw on a simple analogy. Imagine a man wants to build a hill on a flat piece of land. The hill will represent the universe. To make this hill he digs a hole in the ground and uses that soil to dig his hill. But of course he's not just making a hill – he's also making a hole, in effect a negative version of the hill. The stuff that was in the hole has now become the hill, so it all perfectly balances out. This is the principle behind what happened at the beginning of the universe.

When the Big Bang produced a massive amount of positive energy, it simultaneously produced the same amount of negative energy. In this way, the positive and the negative add up to zero, always. It's another law of nature.

So where is all this negative energy

today? It's in the third ingredient in our cosmic cookbook: it's in space. This may sound odd, but according to the law of nature concerning gravity and motion – laws that are among the oldest in science – space itself is a vast store of negative energy. Enough to ensure that everything adds up to zero.

I'll admit that, unless mathematics is your thing, this is hard to grasp, but it's true. The endless web of billions upon billions of galaxies, each pulling on each other by the force of gravity, acts like a giant storage device. The universe is like an enormous battery storing negative energy. The positive side of things – the mass and energy we see today – is like the hill. The corresponding hole, or negative side of things, is spread throughout space.

So what does this mean in our quest

to find out if there is a God? It means that if the universe adds up to nothing, then you don't need a God to create it. The universe is the ultimate free lunch.

Since we know that the positive and the negative add up to zero, all we need to do now is to work out what – or dare I say who – triggered the whole process in the first place. What could cause the spontaneous appearance of a universe? At first, it seems a baffling problem – after all, in our daily lives things don't just materialise out of the blue. You can't just click your fingers and summon up a cup of coffee when you feel like one. You have to make it out of other stuff like coffee beans, water and perhaps some milk and sugar. But travel down into this coffee cup – through the milk particles, down to the atomic level and right down to the sub-atomic level, and you enter a world where conjuring something out of nothing is possible. At least, for a short while. That's because, at this scale, particles such as protons behave according to the laws of nature we call quantum mechanics. And they really can appear at random, stick around for a while and then vanish again, to reappear somewhere else.

Since we know the universe itself was once very small – perhaps smaller than a proton – this means something quite remarkable. It means the universe itself, in all its mind-boggling vastness and complexity, could simply have popped into existence without violating the known laws of nature. From that moment on, vast amounts of energy were released as space itself expanded – a place to store all the negative energy needed to balance the books. But of course the critical question is raised again: did God create the quantum laws that allowed the Big Bang to occur? In a nutshell, do we need a God to set it up so that the Big Bang could bang? I have no desire to offend anyone of faith, but I think science has a more compelling explanation than a divine creator.

Our everyday experience makes us think that everything that happens must be caused by something that occurred earlier in time, so it's natural for us to think that something – maybe God – must have caused the universe to come into existence. But when we're talking about the universe as a whole, that isn't necessarily so. Let me explain. Imagine a river, flowing down a mountainside. What caused the river? Well, perhaps the rain that fell earlier in the mountains. But then, what caused the rain? A good answer would be the Sun, that shone down on the ocean and lifted water vapour up into the sky and made clouds. Okay, so what caused the Sun to shine? Well, if we look inside we see the process known as fusion, in which hydrogen atoms join to form helium, releasing vast quantities of energy in the process. So far so good. Where does the hydrogen come from? Answer: the Big Bang. But here's the crucial bit. The laws of nature itself tell us that not only could the universe have popped into existence without any assistance, like a proton, and have required nothing in terms of energy, but also that it is possible that nothing caused the Big Bang. Nothing.

The explanation lies back with the theories of Einstein, and his insights into how space and time in the universe are fundamentally intertwined. Something very wonderful happened to time at the instant of the Big Bang. Time itself began.

To understand this mind-boggling idea, consider a black hole floating in space. A typical black hole is a star so massive that it has collapsed in on itself. It's so massive that not even light can escape its gravity, which is why it's almost perfectly black. Its gravitational pull is so powerful, it warps and distorts not only light but also time. To see how, imagine a clock is being sucked into it. As the clock gets closer and closer to the black hole, it begins to get slower and slower. Time itself begins to slow down. Now imagine the clock as it enters the black hole – well, assuming of course that it could withstand the extreme gravitational forces – it would actually stop. It stops not because it is broken, but because inside the black hole time itself doesn't exist. And that's exactly what happened at the start of the universe.

In the last hundred years, we have made spectacular advances in our understanding of the universe. We now know the laws that govern what happens in all but the most extreme conditions, like the origin of the universe, or black holes. The role played by time at the beginning of the universe is, I believe, the final key to removing the need for a grand designer and revealing how the universe created itself.

As we travel back in time towards the moment of the Big Bang, the universe gets smaller and smaller and smaller, until it finally comes to a point where the whole universe is a space so small that it is in effect a single infinitesimally small, infinitesimally dense black hole. And just as with modern-day black holes, floating around in space, the laws of nature dictate something quite extraordinary. They tell us that here too time itself must come to a stop. You can't get to a time before the Big Bang because there was no time before the Big Bang. We have finally found something that doesn't have a cause, because there was no time for a cause to exist in. For me this means that there is no possibility of a creator, because there is no time for a creator to have existed in.

People want answers to the big questions, like why we are here. They don't expect the answers to be easy, so they are prepared to struggle a bit. When people ask me if a God created the universe, I tell them that the question itself makes no sense. Time didn't exist before the Big Bang so there is no time for God to make the universe in. It's like asking for directions to the edge of the Earth – the Earth is a sphere that doesn't have an edge, so looking for it is a futile exercise.

Do I have faith? We are each free to believe what we want, and it's my view that the simplest explanation is that there is no God. No one created the universe and no one directs our fate. This leads me to a profound realisation: there is probably no heaven and afterlife either. I think belief in an afterlife is just wishful thinking. There is no reliable evidence for it, and it flies in the face of everything we know in science. I think that when we die we return to dust. But there's a sense in which we live on, in our influence, and in our genes that we pass on to our children. We have this one life to appreciate the grand design of the universe, and for that I am extremely grateful.